Ulisses Gabriel Moraes Lobo
Carina Faria

Grain size, dry matter and diet consumption for Senepol cattle

Ulisses Gabriel Moraes Lobo
Carina Faria

Grain size, dry matter and diet consumption for Senepol cattle

ScienciaScripts

Cover image: www.ingimage.com

This book is a translation from the original published under ISBN 978-613-9-63499-6.

Publisher:
Sciencia Scripts
is a trademark of
Dodo Books Indian Ocean Ltd. and OmniScriptum S.R.L publishing group

120 High Road, East Finchley, London, N2 9ED, United Kingdom
Str. Armeneasca 28/1, office 1, Chisinau MD-2012, Republic of Moldova, Europe
Printed at: see last page
ISBN: 978-620-7-66820-5

SUMMARY

The objective was to evaluate the physical quality of the diet provided throughout the day and its influence on the consumption of Senepol animals participating in a feed efficiency test. The experiment was conducted at the Capim Branco Farm of the Federal University of Uberlândia, using the diet provided to 38 young bulls of the Senepol breed, lasting 91 days, with 21 days of adaptation and 70 days of testing, between January and April of 2015, confined using the *GrowSafe electronic trough system.* The dry matter samples (DM, kg) were processed, consumption data were also collected in the *GrowSafe* system database , such as: total number of visits to the trough (NV, n); number of visits to the trough with consumption (VCC, n); number of visits to the trough without consumption (VSC, n); total consumption in kilograms (CT, kg) and feeding time in minutes (TA, min). The data obtained were subjected to normality tests, after checking the parametric and non-parametric data and mean tests were applied using the SAEG program (2007) and, subsequently, regression and correlation analyses. For the MS variable, there was a significant difference from the first to the third day of collection between times, and there was no difference on the fourth and fifth day of collection. For the other variables (NV, VCC, CT and TA) there was a significant difference between the times on the five days of collection, between these characteristics there was a weak, moderate and strong positive correlation. Linear and quadratic regression analyzes do not explain the behavior of these variables. There was no statistical difference between the times on each day of collection for the granulometry variable, however, between the sieves within each day of collection there was. From this research, it can be concluded that physical changes in the diet throughout the day did not influence the consumption of food from Senepol cattle.

KEYWORDS: *bos taurus.* beef cattle, feeding behavior, diet quality, particle size.

SUMMARY

CHAPTER 1

INTRODUCTION

Beef cattle farming in Brazil is economically consolidated among the main and most important agricultural activities (CEPEA, 2016), being the country with the largest commercial herd in the world and the largest exporter of beef (CARVALHO and ZEN, 2017).

In the last two decades, intensive livestock production systems, called confinement and semi-confinement, have grown and gained space, reaching high levels of productivity, mainly due to the constant use of new cutting-edge technologies within properties, for production and also management of the activity. (CARVALHO and ZEN, 2017).

Combined with the increase in productivity through the use of technologies, we have the insertion of genetic improvement of herds, seeking animals that are increasingly efficient in meat production and, preferably, that consume less food.

Efficient animal selection programs are present in several properties and breeding herds of different bovine breeds, with food efficiency characteristics being the most studied recently, as they are directly related to the higher costs in beef production. Animals that use ingested food more efficiently need to consume less food to reach the same production levels as others that consume more, thus becoming more profitable and producing more meat per unit area (MENDES and CAMPOS, 2016) .

When animals are subjected to performance tests in groups of contemporaries, in order to make comparisons fair, phenotypic factors of behavior and group hierarchy can influence this characteristic, especially when the batches for comparisons are formed shortly before the start of the tests.

Therefore, there is a need for time to establish hierarchy in the group (QUINTILIANO and COSTA, 2006). We also have the fact that after dominance is established, subordinate animals will always have to wait for the dominant animals to eat and then feed themselves.

Cattle are animals with a selective diurnal eating habit (CUSTODIO, 2016), which have the ability to select food for ingestion, a fact that interferes with the nutritional composition of the food consumed and trough leftovers, being able to privilege and maximize the performance of which they feed. first, and harm those who eat last, even leading to metabolic disorders. When confined and fed in troughs, the ration must be divided throughout the day and the quality of the mixture must provide the same ration for everyone, and there must always be food availability, to try to minimize this selection effect (CUSTODIO et al., 2016).

In order to control the quality of a diet throughout the day, some simple methods can be applied to predict this information. Measuring the dry matter and granulometry of ingredients and feed are simple activities that should become routine and can explain changes in animal behavior and performance. Drastic variations in these variables imply fluctuations in the concentrations of nutrients in the food available in the trough, in the density of the diet and even in fermentation, oxidation and degradability in the trough and rumen (PEREIRA, 2016).

Due to the dietary and behavioral factors that can influence the results of feed efficiency tests for beef cattle, for the selection of superior animals, it is essential to know the changes in the diet, as well as the granulometry throughout the day, in order to to obtain subsidies that help to predict with greater accuracy the results achieved in food efficiency assessments to obtain CAR measurements.

Thus, the objective was to evaluate the diet in its physical composition, throughout the day, and its influence on the consumption of Senepol animals participating in feed efficiency tests.

CHAPTER 2

THEORETICAL REFERENCE

2.1 Genetic parameters for consumption

The analyzes of genetic parameters are essential for the establishment of guidelines that guide future actions in beef cattle herds, with studies on improving zebu cattle being scarce, due to the vast majority of commercial herds of global importance being composed of taurine cattle (EUCLIDES FILHO , 2009).

There are many publications with estimates of genetic parameters of feed efficiency traits in beef cattle, especially taurine breeds *(Bos taurus).* Moraes et al., (2016) carried out a survey of studies carried out in the last 10 years that address genetic parameters (heritability, genetic correlation and additive genetic variance) for consumption characteristics and feed efficiency in cattle, presenting estimates of the application of this characteristic in breeding genetics of different breeds (Table 1).

As shown in table 1, it is observed that for the characteristics that were heritability measured, for CAR (Residual Food Intake) most of the values are of average classification, for high consumption and GPD (Daily Weight Gain) also average. As for genetic correlations between CAR and Consumption, most of the results are of moderate correlation and between CAR and GPD, most of the values are weak, with nulls and also negatives.

The characteristic additive genetic variance is expressed in table 1 with mostly low results, with many of the studies also lacking values, a fact that can be attributed to being more difficult to measure and taking longer. However, the V CAR results must be interpreted according to the fact that the genetic gain is dependent on the heritability of the evaluated characteristic, the intensity of selection practiced in the herd and the variance of the characteristic, when selection is carried out for a long time with a certain focus, the tendency is for the genetic variance of the herd to decrease and consequently heritability will be low (PEREIRA, 2014).

Table 1 - Heritability estimates, genetic correlations and additive genetic variation of beef cattle of different breeds.

Flock	AT No.	Heritability			Genetic correlations		VCAR
		CAR	Cons	GPD	CARx Cons	CARx GPD	
Wagyu	740	0.24	0.78	0.25	0.78	0.25	-
Wagyu	1304	0.29	0.33	0.26	0.56	-0.14	0.41
Angus X Charolais	813	0.21	0.54	0.59	0.73	0.46	0.15
Angus X Charolais	490	0.42	0.34	0.23	0.59	-0.15	-
Angus X Charolais	721	0.29	0.41	0.28	-	-	-
Brahman	468	0.47	0.48	0.21	0.59	0.02	0.19
Composite Tropical	1209	0.31	0.51	0.50	-	-	-
Brangus	468	0.47	0.48	0.21	0.85	0.04	-
Angus	862	0.14	0.14	0.09	0.58	0.00	-
Dutch	903	0.27	0.17	0.22	-	-	0.44
Irish	2605	0.45	0.49	0.30	0.59	0.01	0.11
Charolais	417	0.68	0.54	0.54	0.52	-0.01	-
Nellore	678	0.33	0.60	0.42	0.33	0.06	0.04
Nellore	1038	0.37	0.40	0.35	0.05	0.05	0.20

AT No. = Number of animals tested
CAR = Residual Food Intake (kg/day)
Cons = Consumption (kg/day)
GPD = Daily Weight Gain (kg/day)
V CAR = Additive genetic variance for CAR (kg/day)
Source: Adapted from Moraes et al., (2016)

It is extremely difficult for a given genotype to have all the desired adjectives inserted within a careful selection process, and the use of indices, with different weights for certain characteristics, facilitates the process of identifying superior animals, therefore the animals must be continually tested. and evaluated to improve the desired characteristics (MENDES and CAMPOS, 2016).

2.2 Food efficiency tests

They consist of experiments conducted with animals, in this case cattle, at similar age ranges and body weights (groups of contemporaries), with the aim of selecting the best animals for a given pre-established characteristic. Age intervals greater than 60 days make comparisons unfair, as they cause changes in the maintenance requirement. Variations in efficiency may be the result of animals evaluated under different environmental conditions, gain composition and food consumption (CASTILHOS, 2009).

Measuring individual consumption is a process that requires specific infrastructure and labor to collect data, mainly on the food provided and leftovers. Individual bay systems and manual data collection are restricted to a few, due to the system infrastructure, required labor and installation purchase price (MENDES and CAMPOS, 2016).

Some basic criteria must be followed to carry out these tests, such as: age interval between animals of a maximum of 60 days, similar body weight, similar previous feeding conditions, animals between 8 and 33 months of age. If the animals come from different properties, there is a minimum period of adaptation to the test site (paddock or pasture) and to other animals, in case of confinement, adaptation to the diet of at least 21 days, weekly feed collections to check quality (at least from MS). It is also recommended that the diet be standardized throughout the course of the test, provided at will, at least twice a day (GOMES and MENDES, 2013).

The inclusion of food efficiency as a tool in a selection program, consequently, inserts evidence of efficiency, if planned in the long term, due to the benefits it can bring, it is economically viable, once the structure is established, most of the other costs They are not fixed (food, labor, among others), and can carry out tests on a herd for many years. Increasing the herd's feeding efficiency is a way to increase profitability by reducing food costs, of course this must be thought about in the long term. Furthermore, studies prove the reduction in greenhouse gas emissions, with an important conservation appeal (OLIVEIRA, 2013).

2.3 Cattle feeding behavior

When batches of animals are formed, regardless of the objective, they will undergo the establishment of a hierarchy within the group (QUINTILIANO and COSTA, 2006). In groups of contemporaries, this is no different, requiring a few weeks to establish the dominant (they feed first) and dominated (they feed last) animals, as well as the new environment.

The association of the duration of feeding time in the trough, with the times at which the animals preferably feed, is important for establishing appropriate management strategies for each situation. Variations in food consumption and feeding frequency can be evidenced in the assessment of eating behavior, which has long been done through visual assessments, and there are currently specific feeding systems to measure such characteristics (PAZDIORA et al., 2011; NKRUMAH et al., 2006).

By studying feeding behavior, it is possible to evaluate some characteristics of biological and economic importance, such as feeding time, expressed as the time the animal spent ingesting food; feeding frequency, which reveals the number of visits the animal makes to the trough daily; feeding rate or efficiency, calculated as the amount of food consumed per visit, among others (LAGE, 2013).

Silva (2014a) found in his study that animals fed more than twice a day have a greater consumption of dry matter, and less variation in this characteristic, when compared to animals fed once or twice a day. Individuals who were fed more times a day also selected less food and spent more time ruminating , a fact that maintains the initial composition of the feed provided for a longer period.

The levels of roughage in a diet influence the consumption and performance of animals, due to the physical aspects of regulating consumption, which limit selection, ingestion, and degradation after a certain point. Detmann et al (2003), recommend that, for animals to have standardization in the composition of feed and consumption, the physical quality of the mixture and ingredients must be known and always monitored.

2.4 *GrowSafe* feeding system

The GrowSafe System® is characterized by an electronic system for monitoring feed intake for cattle, the company owns it of Canadian origin, uses radio frequency technology and allows individual feeding data to be documented, with a degree of sensitivity that has never been achieved by observation visual. The system was designed to record visits to each animal's trough, individually, by reading ear tags (SCHWARTZKOPF-GENSWEIN et al., 2002).

Among the main advantages of GrowSafe, emphasis is placed on the use of several troughs deployed in large areas, being able to measure the individual consumption of all animals in the flock. This system has advantages over other existing ones such as the "calan gate" system or individual stalls, as it continuously measures food consumption, in addition to the frequency of each animal's visit to the feeder and duration (in seconds) of feeding (VALADARES FILHO et al ., 2006).

They also include other important observations per visit to the trough, such as: interval between visits to the trough; animals remaining in the trough but not consuming; number of animals feeding simultaneously (if any) and the order in which the animals feed, which characterizes dominant and dominated animals (MEDEIROS et al., 2016).

Medeiros et al. (2016) cite as the biggest advantage of this system that feeding is done in groups, with the possibility of a large number of animals per batch, with no need for training, and also emphasizes the possibility of measuring IMS continuously every second in the trough, with accuracy of 10 grams.

2.5 Diet particle size analysis

A practice commonly used in beef cattle confinements is feeding in the form of total ration, which consists of mixing the roughage and concentrated fraction of the diet, feeding it together. However, there is concern about this type of feeding, due to the natural habit of animals selecting, for or against, certain components of the diet according to particle size (CUSTODIO et al., 2016).

According to Custodio et al. (2016), studies on the selective consumption of feed particles in confined beef cattle are scarce, but there are consistent reports in the literature on the selectivity and food preference of confined dairy cows, which express a preference for concentrate.

Krause and Oetzel (2006), studying the feeding behavior of dairy cows, found that the selection carried out in the ration by the dominant cows resulted in reducing the nutritional value of the ration that was available to the dominated cows.

Devriets et al. (2008), state that, when cows select food, and give preference to ingesting the concentrate, rejecting the roughage fraction, nutritional needs can be exceeded, causing undesirable rumen fermentation, causing metabolic disorders such as acidosis and bloating and consequently pathologies.

Based on the need to know the selectivity of feed ingredients by animals through a particle separator, equipment called *Penn State Particle Separators* (SPPS) was developed, which can evaluate total feed, *fresh forage,* as well as silage and pre-dried (STIEVEN, 2012; LUIZ, 2014).

The SPPSs are made up of a set of four containers, three of which are sieves, which reduce the size of the holes in millimeters so that the longest particles are in the first, medium particles in the second, short particles in the third and the fourth and the last container (bottom of the sieve) contains the fine particles (SILVA, 2014b).

Understanding the mechanisms that influence the selection of particles in the diet of confined beef cattle becomes useful to optimize the performance of the animals, and it is essential to choose the best feeding strategy according to each situation and management (SILVA, 2014a).

2.6 Behavior and fractionation of the diet in confinement

Total rations with the inclusion of bulky ingredients and/or low dry matter contents may present problems with food fermentation and loss of food quality throughout the day, generating a drop in consumption and, consequently, a decrease in performance (TAKIGAWA , 2015).

Researchers mention that according to the increase in the number of daily treatments, the performance of confined cattle can be improved, due to greater incentive for the animals to feed and also the renewal of feed in the trough (DAMASCENO et al., 1999; SCHWARTZKOPF- GENSWEIN et al., 2003; SCHUTZetal., 2011).

Rotger et al. (2006) found results where a greater fractionation of the diet throughout the day does not differ in the animals' performance, and Silva (2014a) concluded that a greater fractionation of the diet can increase the number of metabolic disorders in animals, mainly due to high intake of concentrate.

Silva (2014b) reports that there are several points that must be taken into consideration when choosing nutritional and food management, such as particle size of the food that will make up the total ration, methods and place of feeding, prior or late contact with food available for ingestion, batch size and homogeneity, space per animal, among others, and tests must be carried out to clarify how and to what extent these factors affect production.

2.7 Relationship of consumption with diet quality and animal performance

Every confinement feed can have up to four different compositions: the formulated feed, the feed mixed for supply, the feed offered in the trough, and the feed consumed. Being the latter, it must have nutritional values and bromatological composition very close to the values of the former, thus ensuring the animals' access to nutrients in quantity and quality, as they need according to the formulation (LAZARINI et al., 2014; COSTA JÚNIOR et al., 2016).

A good mixture of the total feed, if ingredients with ideal quality and particle size are used, which promotes good adhesion between the particles of different ingredients, provides less selection of the different portions that make up the feed by the animals (especially in high-concentrate diets). Cattle with free access to the trough and feed generally separate the ingredients by particle size and density and consume the components of their preference (GOULART, 2012).

Therefore, diets must be provided in the most standardized and uniform way possible, and the mixing time of the wagon and/or mixer (according to each manufacturer's recommendation) must be respected, so that the composition of the feed supplied at the beginning of the trough line, is very close to that supplied at the end of the trough line, thus ensuring the supply and access of all animals to the same feed (SOVA et al., 2014).

Regardless of high or low intakes, or the batch's feeding hierarchy, the quality and standardization of the diet is a factor that can directly influence the animals' weight performance and feeding efficiency, and can also minimize the occurrence of metabolic disorders. Therefore, attention must be paid to the quality of the mixing of ingredients, as this can affect the

meeting nutritional requirements and consequently the final performance of cattle in a confinement system (MARANO et al., 2014; GROESBECK et al, 2006).

CHAPTER 3

METHODOLOGY

The experiment was conducted at the Federal University of Uberlândia (UFU), at the Capim Branco Experimental Farm, in Uberlândia - MG.

We used the dietary information provided for 38 young bulls with an average age of 15 months, not castrated, participating in the Senepol Bull Feeding Efficiency Test, lasting 91 days, with the first 21 days of adaptation, between January and April. from 2015, coming from 15 rural properties, confined using the *GrowSafe electronic system and ad libitum* diet , aiming for average daily gains of 1.3kg/animal/day. The diet was formulated in a 60:40 proportion of bulky corn silage and concentrate based on ground corn, soybean meal, urea and mineral core (Table 2), meeting NRC parameters (2000) and feeding three times a day (8:3 Oh, 10:30 am and 3:30 pm) with adjustment daily, so that there were leftovers of around 10% of what was supplied, to guarantee the measurements of the electronic system of the troughs and food for the animals throughout the night, until the early morning of the next day, which were removed every day in the morning, before the first treatment.

Table 2 - Inclusion of rations used during the feed efficiency test.

	Inclusion of diets		
Ingredients	Adaptation I	Adaptation II	Termination
Silage	84.65%	73.30%	60.00%
Ground corn	8.00%	15.00%	28.00%
Soybean meal	4.00%	8.00%	8.00%
Urea	0.35%	0.70%	1.00%
mineral core	3.00%	3.00%	3.00%
GMD (kg)	0.850	1,000	1,300
Period (days)	11	10	70

The animals evaluated in the efficiency test were kept in confinement, in an area of 1,680 m^2 (42 m x 40 m), divided into two corrals, where the animals were housed randomly, for 91 days, with weighing every 14 days, with a 2,600-liter central drinking fountain in the paddocks, and four *GrowSafe System electronic troughs* per corral (a total of 8 troughs, covered by asbestos tiles), and according to the manufacturer's recommendation, one trough can feed up to 8 animals efficiently and safe.

Collections of total feed to measure dry matter (DM) and granulometry were carried out every 14 days, always one day before the date of weighing the animals, so that there was no interference with consumption and eating behavior, totaling six days of collection (02/15, 03/01, 03/15, 03/29, 04/12 and 04/26) within the termination period.

For MS, collections were made in 5 locations in each trough, in a representative way, then homogenization was carried out and a sample was removed, which was taken to the laboratory for measurement using a forced ventilation oven. For granulometry, the collection method was the same, however, the final sample taken of approximately 500 g was processed and placed in the PSPS *in natura.* Each trough was represented by a sample in each period and at the time of collection, the animal(s) that were in the trough were removed.

Penn State Particle Separator (PSPS) was used , which is composed of three sieves with holes measuring 19, 8, 1 millimeters (mm), respectively, from above. downwards and a tray at the bottom without holes, according to the methodology proposed by Lopes (2011), which recommends for each sample to move from right to left for approximately one second (Is) and left to right (Is) for 5 times (Figure 1). Repeat 8 times (8 beats), totaling 40 times. After separating the particles, weigh each container.

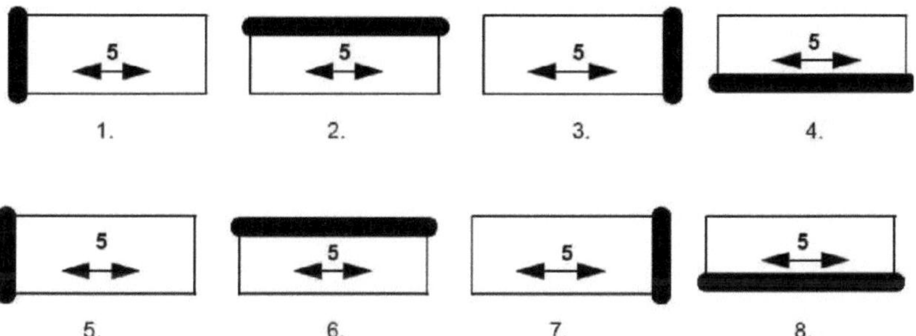

Penn State Particle Separator (PSPS) agitation mode for particle separation on sieves.

For the two variables mentioned (DM and granulometry), 14 collections were made, in DIC, between 8:30 am and 8 pm (time of greatest activity of the animals), with a time interval of 30 to 60 minutes (Table 3), the which represent the treatments (Time 1 to 14) and the troughs, the repetitions (8 troughs) in this experiment. The shorter time intervals (30 minutes) were used to investigate the increase in cattle behavior at times close to feeding.

Table 3 - Total diet collection times for DM and granulometry.

Time	Time (h)
1	08:30
two	09:00
3	10:00 am
4	10:30
5	11:00
6	12:30
7	13:30
8	14:30
9	15:30
10	4:00 pm
11	17:00
12	18:00
13	19:00
14	8:00 pm

data on eating and consumption behavior were collected in the *GrowSafe* system database , such as: total number of visits to the trough (NV, n); number of visits to the trough with consumption (VCC, n); number of visits to the trough without consumption (VSC, n); total consumption in kilograms (CT, kg) and feeding time in minutes (TA, min). To collect these variables, time intervals were used between each collection of the total diet (Table 4), given that, on each day of sampling and measurements, there may have been different factors that could influence the results.

Table 4 - Time interval used to calculate the variables NV, VCC, VSC, CT and TA.

Time	Time interval (h)
1	8:30 to 9:00
two	9:00 to 10:00
3	10:00 to 10:30
4	10:30 to 11:00

5	11:00 to 12:00
6	12:00 to 12:30
7	12:30 to 13:30
8	13:30 to 14:30
9	2:30 pm to 3:30 pm
10	15:30 to 16:00
11	16:00 to 17:00
12	17:00 to 18:00
13	18:00 to 19:00
14	7:00 pm to 8:00 pm

The data obtained from the diet and food consumption were subjected to normality tests, after checking the parametric and non-parametric data and mean tests were applied using the SAEG program (2007), with Kruskal-Wallis for the non-parametric variables and Scott -knott for parametric variables, both at 5% probability.

Linear and quadratic regression analyzes were also carried out for parametric variables and Pearson correlation for consumption characteristics and eating behavior.

The data from April 12th was removed from the evaluations due to typing errors in the values.

CHAPTER 4

RESULTS AND DISCUSSION

For the variables NV, VCC, TA, MS and CT, the results are described in Tables 5 to 9, with each table representing a collection day (15/02, 01/03, 15/03, 29/03 and 26/04), totaling five collections.

Table 5 - Mean characteristics of total number of visits (NV, n), number of visits with consumption (VCC, n), feeding time (TA, min), dry matter (DM, %) and total consumption (TC, kg) in times 1 to 14 on the first day of collections.

			First Collection (02/15)		
Time	NV1	VCC1	TA1 (Min)	MSI (%)	CT1 (kg)
1 (8:30 to 9:00)	10.12 ^{ab}	9.62 ^{ABC}	28.06 ^{bc}	42.3 ^{lb}	6.90c
2 (9:00 to 10:00)	15.00 ^{ab}	14.75 ^{to}	54.56 ^{a}	42.25b	11.36 ^{a}
3 (10:00 to 10:30)	7.37 ^{ABC}	6.87 ^{ABC}	25.78 ^{bc}	41.74b	5.72c
4 (10:30 to 11:00)	8.00 ^{ABC}	7.75 ^{ABC}	26.05 ^{bc}	41.99b	5.48c
5 (11:00 to 12:00)	11.87 ^{ab}	11.37 ^{ab}	62.54 ^{to}	40.96'	10.85 ^{to}
6 (12:00 to 12:30)	1.75c	1.75c	18.72c	42.24 ^{b}	2.35d
7 (12:30 to 13:30)	11.75 ^{ab}	10.87 ^{ab}	50.39 ^{ab}	40.18c	8.55b
8 (13:30 to 14:30)	14.25 ^{ab}	12.87 ^{a}	51.40 ^{ab}	42.02b	7.80b
9 (2:30 pm to 3:30 pm)	11.00 ^{bc}	9.87 ^{ab}	60.93 ^{ab}	43.80 ^{to}	8.22c
10 (15:30 to 16:00)	4.62 ^{bc}	4.25 ^{bc}	33.24 ^{ABC}	42.27b	6.30c
11 (4:00 pm to 5:00 pm)	7.50 ^{ABC}	7.25 ^{ab}	61.92 ^{a}	45.44 ^{to}	6.51c
12 (5:00 pm to 6:00 pm)	7.75 ^{ABC}	7.00 ^{ABC}	52.02 ^{ab}	44.02 ^{to}	7.71 ^{b}
13 (6:00 pm to 7:00 pm)	11.50 ^{ab}	11.00 ^{ab}	52.05 ^{ab}	42.18b	7.71 ^{b}
14 (7:00 pm to 8:00 pm)	10.12 ^{ab}	9.87 ^{ABC}	43.50 ^{ABC}	43.00b	6.80b
s	3.60	3.41	15.06	1.30	2.23
CV	0.38	0.38	0.34	0.03	0.30

*different letters in the same column differ from each other using the Kruskal-Wallis test at 5% probability for the variables NV, VCC and TA.

*different letters in the columns of the MS and CT variables differ from each other using the Scott-knott test at 5% probability

S = standard deviation from the mean

CV = Coefficient of variation

Table 6 - Average characteristics of total number of visits (NV, n), number of visits with consumption (VCC, n), feeding time (TA, min), dry matter (DM, %) and total consumption (TC, kg) in times 1 to 14 on the second day of collections.

			Second Collection (01/03)		
Time	NV2	VCC2	TA2 (Min)	MS2 (%)	CT2 (kg)

1 (8:30 to 9:00)	5.50 ^{b}	5.50 ^{b}	18.35b	46.87 ^{a}	3.56d
2 (9:00 to 10:00)	13.50 ^{to}	13.38 ^{a}	49.90 ^{to}	42.09c	9.59 ^{to}
3 (10:00 to 10:30)	5.25b	5.25b	23.27b	41.92c	3.92d
4(10:30 to 11:00)	6.13	6.13 ^{ab}	26.97b	42.17c	5.70c
5(11:00 to 12:00)	10.88 ^{ab}	10.88 ^{ab}	51.23 ^{a}	44.20 ^{b}	10.47 ^{a}
6 (12:00 to 12:30)	7.13 ^{ab}	7.00 ^{ab}	24.60b	38.90d	4.21d
7 (12:30 to 13:30)	10.38 ^{ab}	10.25 ^{ab}	49.14 ^{a}	38.30 ^{d}	8.94 ^{to}
8 (13:30 to 14:30)	8.75 ^{ab}	8.62 ^{ab}	43.29 ^{a}	38.13d	7.49b
9 (2:30 pm to 3:30 pm)	6.37 ^{ab}	6.25 ^{ab}	28.38 ^{ab}	47.82 ^{to}	4.86c
10 (15:30 to 16:00)	4.25b	4.00 ^{b}	24.76b	45.84 ^{to}	5.13c
11 (4:00 pm to 5:00 pm)	13.14 ^{ab}	13.00 ^{ab}	44.56 ^{a}	47.53 ^{to}	10.62 ^{a}
12 (5:00 pm to 6:00 pm)	9.50 ^{ab}	9.12 ^{ab}	52.86 ^{to}	43.60b	10.38 ^{to}
13 (6:00 pm to 7:00 pm)	8.37 ^{ab}	8.00 ^{ab}	41.68 ^{ab}	41.77c	7.14 ^{b}
14 (7:00 pm to 8:00 pm)	10.00 ^{ab}	9.75 ^{ab}	16.20b	37.71d	3.14d
s	2.89	2.87	13.29	3.52	2.78
CV	0.34	0.34	0.38	0.08	0.41

*different letters in the same column differ from each other using the Kruskal-Wallis test at 5% probability for the variables NV, VCC and TA.

*different letters in the columns of the MS and CT variables differ from each other using the Scott-knott test at 5% probability

S = standard deviation from the mean

CV = Coefficient of variation

In tables 5 and 6, the lowest values of the variables NV, VCC, TA and CT are observed in the five times where the measurement interval was 30 minutes (times 1, 3, 4, 6 and 10), however when considering statistical significance, the shortest times were not the least representative, a fact that can be explained by greater activity in the animals' trough, as they were close to feeding times. The variation in consumption was not related to the variation in DM.

Table 7 - Mean characteristics of total number of visits (NV, n), number of visits with consumption (VCC, n), feeding time (TA, min), dry matter (DM, %) and total consumption (TC, kg) in times 1 to 14 on the third day of collections.

			Third Collection (15/03)		
Time	NV3	VCC3	TA3 (Min)	MS3 (%)	CT3 (kg)
1 (8:30 to 9:00)	10.37 ^{ab}	10.00 ^{ab}	25.11 ^{ab}	38.24 ^{a}	8.33c
2 (9:00 to 10:00)	25.38 ^{a}	24.75 ^{a}	52.90 ^{to}	37.87 ^{a}	14, 52 ^{a}
3 (10:00 to 10:30)	7.12 ^{ab}	7.00 ^{ab}	20.35b	38.80 ^{to}	4.52d
4 (10:30 to 11:00)	5.37b	5.12 ^{b}	24.20b	40.26 ^{to}	6.59c
5 (11:00 to 12:00)	17.88 ^{a}	17.62 ^{a}	54.33 ^{to}	38.38 ^{to}	13.91 ^{to}
6 (12:00 to 12:30)	6.75 ^{ab}	6.63 ^{ab}	26.08 ^{ab}	36.84 ^{to}	5.81d
7 (12:30 to 13:30)	12.62 ^{ab}	12.25 ^{ab}	49.37 ^{a}	33.83b	10.71b
8 (13:30 to 14:30)	8.00 ^{ab}	7.75 ^{ab}	38.53 ^{ab}	38.78 ^{a}	7.27c
9 (2:30 pm to 3:30 pm)	8.88 ^{ab}	8.50 ^{ab}	29.23 ^{ab}	39.41 ^{a}	5.12d
10 (15:30 to 16:00)	6.37b	6.38b	21.73b	37.83 ^{to}	5.54 ^{d}

11 (4:00 pm to 5:00 pm)	11.00 ab	10.63 ab	48.59 a	39.45 to	1 1.05 b
12 (5:00 pm to 6:00 pm)	7.50 ab	7.50 ab	46.30 to	-	10.36 b
13 (6:00 pm to 7:00 pm)	8.75 ab	8.37 ab	31.90 ab	-	7.15c
14 (7:00 pm to 8:00 pm)	5.13b	5.12 b	16.61c	-	3.16d
s	5.52	5.39	13.27	1.71	3.08
CV	0.55	0.55	0.38	0.04	0.40

worse food supply at the end of the day, with all animals having the opportunity to consume standardized, quality food.

Table 8 - Mean characteristics of total number of visits (NV, n), number of visits with consumption (VCC, n), feeding time (TA, min), dry matter (DM, %) and total consumption (TC, kg) in times 1 to 14 on the fourth day of collections.

		Fourth Collection (03/29)			
Time	NV4	VCC4	TA4 (Min)	MS4 (%)	CT4 (kg)
1 (8:30 to 9:00)	5.88 ab	5.88 ab	28.08a bc	37.76	7.3 8b
2 (9:00 to 10:00)	7.87 to	7.87 to	47.93 to	33.76	10.35 to
3 (10:00 to 10:30)	5.13 ab	4.88 ab	22.47 ABC	37.58	5.66b
4(10:30 to 11:00)	5.13 ab	5.13 ab	22.76 ABC	41.48	5.54b
5(11:00 to 12:00)	9.25 to	9.25 to	51.16 a	39.56	11.44 a
6 (12:00 to 12:30)	1.63b	1.63b	17.28bc	39.96	2.77c
7 (12:30 to 13:30)	5.38 ab	5.25 ab	32.51 ab	39.19	5.73b
8 (13:30 to 14:30)	4.50 ab	4.50 ab	23.15 ABC	38.24	4.19c
9 (2:30 pm to 3:30 pm)	2.25 ab	2.13 ab	13.81 bc	40.02	2.83c
10 (15:30 to 16:00)	2.75 ab	2.63 ab	20.90 ABC	39.32	5.81b
11 (4:00 pm to 5:00 pm)	10.00 to	10.00 to	48.86 to	38.95	12.23 to
12 (5:00 pm to 6:00 pm)	9.50 to	9.38 to	50.10 to	38.92	13.08 to
13 (6:00 pm to 7:00 pm)	4.75 ab	9.75 ab	25.03 ABC	38.06	6.10b
14 (7:00 pm to 8:00 pm)	1.50b	1.38b	5.93c	36.04	1.21c
s	2.87	3,H	14.67	1.88	3.71
CV	0.53	0.56	0.50	0.05	0.55

Table 9 - Average characteristics of total number of visits (NV, n), number of visits with consumption (VCC, n), feeding time (TA, min), dry matter (DM, %) and total consumption (TC, kg) in times 1 to 14 on the fifth day of collections.

		Fifth Collection (04/26)			
Time	LV5	VCC5	TA5 (Min)	MS5 (%)	CT5 (kg)
1 (8:30 to 9:00)	4.3 8 b	4.38b	19.64 ABC	38.14	4.4 1 b
2 (9:00 to 10:00)	10.63 to	10.63 to	31.12 ABC	36.34	7.63 to
3 (10:00 to 10:30)	5.38 ab	5.25 ab	12.90b	37.03	3,ll c
4 (10:30 to 11:00)	9.13 ab	8.88 ab	21.68 ABC	39.53	5.90b
5 (11:00 to 12:00)	10.88 to	10.75 to	36.68 a	43.48	7.87 to
6 (12:00 to 12:30)	7.13 ab	7.13 ab	20.66 ABC	38.86	45.37b
7 (12:30 to 13:30)	8.50 ab	8.50 ab	36.58 a	42.72	8.15 to
8 (13:30 to 14:30)	6.00 ab	6.00 ab	27.58 ABC	38.99	5.45b
9 (2:30 pm to 3:30 pm)	4.3 8 b	4.38b	13.61b	41.00	2.74c
10 (15:30 to 16:00)	7.38 ab	7.38 ab	21.68 ABC	41.74	6.83b

11 (4:00 pm to 5:00 pm)	9.88 ab	9.88 ab	30.08 ABC	39.49	8.57 a
12 (5:00 pm to 6:00 pm)	10.63 to	10.63 to	33.11 ab	37.65	10.12 to
13 (6:00 pm to 7:00 pm)	7.13 ab	7.00 ab	14.48 bc	41.00	4.62b
14 (7:00 pm to 8:00 pm)	5.63 ab	5.63 ab	11, 17c	35.03	3.30c
s	2.32	2.31	8.89	2.44	10.75
CV	0.30	0.30	0.38	0.06	1.21

*different letters in the same column differ from each other using the Kruskal-Wallis test at 5% probability. S = standard deviation from the mean
CV = Coefficient of variation

In tables 8 and 9, the data for the variables NV, VCC, TA and CT showed significant differences between times, depending on the previous evaluation days, with an irregular trend.

On all collection days (Tables 5 to 9), variations in the significance of results can be observed throughout the day, from early morning to sunset, a fact that can be attributed to the maintenance of the standard and quality of the diet. , due to the cover that existed over the troughs. Since the search for food and consumption suffered a trend throughout the day, the animals showed inconstant eating behavior, looking for troughs and feeding at a variable time and quantity over the days, but with greater intensity in times close to the scheduled times. feed supply.

For the MS variable (Tables 5 to 9), the results from the first two days of collection showed significant differences between times, with no tendency to increase or decrease as the hours passed.

As explained by Signoretti et al. (1999), Coelho (2002), Rustomo et al. (2006) and Pereira (2016), after formulating and manufacturing a diet, if it shows a tendency to increase DM throughout the day, this may characterize a loss of quality and standard of the same, due to the selection behavior benefited by factors such as dominance, poor quality of the mixture and particle size of the ingredients, which did not occur in this study. Inconstant variations during the day should be interpreted as normal activity, as factors such as temperature and collection homogenization have a direct influence.

In this experiment, the results show that the food remained constant throughout the day, which allows us to affirm that the diet remained stable and the subordinate animals were not harmed in their performance, so that they could express their feeding efficiency in RFI measurements.

Garcia (2009), recommends measuring DM several times a day, as measured in this study, so that we have concrete information on the behavior of the feed, and if necessary, take appropriate actions according to each situation. Unlike this study, there were large fluctuations in this characteristic in this author's studies with dairy cattle, an undesirable fact due to the feeding and ruminal behavior of the species.

On the third day of data collection (Table 7), samples from the last three times (12, 13 and 14) were lost. The result between the times on this collection day showed little significance, where only the seventh collection of the day presented a lower value than the others, which can possibly be attributed to an error in collection and sampling.

In the last two days of collection (Tables 8 and 9) the DM levels did not differ statistically, a fact that can be explained by the change of silo and, consequently, of the silage, which improved the quality of the diet and kept it homogeneous in the course of the day.

According to Almeida (2006), large changes in DM within roughage silos are frequently found, because in the process of filling and emptying the silo, there can be several errors, mixing forage from different areas, cuts, vegetative stages, rain, among others. . Due to these facts, there may also be impactful changes in the DM of the total ration, which are likely to cause gradual drops in the productivity and health of the herd, as well as waste of inputs, especially when working with high proportions of roughage in the formulations.

In his experiments, Lopes (2011) also carried out MS analyzes to compare with eating behavior later, but only at the conference level three times a day, but concluded that the diet remained stable in terms of this variable.

Brito (2017), comments that for beef cattle, diets with low levels of roughage, when these ingredients are added incorrectly (lower DM levels than in the formulation), can cause metabolic disorders such as acidosis, decreased consumption , which will consequently harm the performance of the animals. In the study in question, the low levels of DM variation are attributed to the fact that the silos are managed appropriately, and the troughs are covered by a masonry structure and tiles.

Results of work to discuss the other characteristics mentioned in this study (NV, VCC, TA and CT), according to the methodology used, throughout the day and through electronic systems, in the literature, to discuss these results, are scarce, in evidence of feed efficiency and the Senopol breed even more inhospitable.

Más, Silva et al (2018), studying the feeding behavior of Nelore cattle in Dracena - SP, visually found the absence of significant effects on feeding frequency, consumption selectivity, time resting and feeding, at finishing, during of the day, in animals that also received food three times a day. However, their evaluations were carried out only twice in the total period of confinement (94 days), for a period of 24 hours, for each animal individually, with 5-minute observations, with intervals not mentioned.

The results of Silva et al (2018) differ from those found in this experiment, because as shown in tables 5 to 9, there were significant differences throughout the day for consumption and behavioral characteristics. However, this fact does not mean that there was a change in the quality of the diet over the days.

According to Rutter (2002), a herd on pasture usually has greater grazing activity at dawn and dusk, at the end of the morning and after dusk most of the animals are ruminating. Under normal circumstances, in the afternoon (after noon) some animals return to grazing, sporadically, with some individuals ruminating or idling, while others graze. In the study in question, the results do not follow this behavior, where the animals' greatest feeding activity was found close to feeding times, therefore this fact had a direct influence.

In grazing situations, cattle meals last an average of 110 minutes and are frequent 5 times a day (Phillips and Rind, 2002). In this study, the food was supplied in troughs, as the animals were in a confinement system and in the form of total feed (roughage plus concentrate), and the results were not in accordance with the study mentioned, in minutes nor in number of times.

As stated by Neumann et al. (2015), carrying out studies on heifers confined for subsequent slaughter, testing diets with different concentrations and sources of roughage, feeding time and behavior was directly influenced by the concentration of roughage in the diet, the more roughage, the longer the feeding time, however , the amount of food ingested in kilograms was not influenced by the type of food. In the study in question, different diets were not tested, but feeding time tended to decrease as the days went by, a fact that can be attributed to the energetic satiety of the cattle, close to the end of the efficiency test.

Pearson correlation analyzes were performed between the characteristics MS, NV, VCC, VSC, CT and TA (Table 10), separated by collection day (1 to 5).

Table 10 - Pearson correlations for the characteristics Dry matter (DM), number of visits (NV), visits with consumption (VCC), visits without consumption (VSC), total consumption (TC) and feeding time (TA), in five days of evaluation (collections 1 to 5).

First-collection correlations						
	MSI	NV1	VCC1	VSC1	CT1	TA1
MSI		-0.01	-0.01	-0.02	-0.13	0.08
NV1	NS	-	0.99	0.45	0.51	0.21
VCC1	NS	***	-	0.33	0.53	0.21
VSC1	NS	***	**		0.10	0.05
CT1	NS	***	***	NS		0.70
TA1	NS	*	*	NS	***	
Seeunda correlations collect						
	MS2	NV2	VCC2	VSC2	CT2	TA2
MS2	-	-0.12	-0.12	-0.04	0.09	-0.02
NV2	NS	-	1.00	0.21	0.56	0.45
VCC2	NS	***		0.11	0.57	0.45
VSC2	NS	NS	NS		0.01	0.07
CT2	NS	***	***	NS	-	0.92
TA2	NS	***	***	NS	***	
Third collection correlations						
	MS3	NV3	VCC3	VSC3	CT3	TA3
MS3		-0.01	-0.02	0.17	-0.01	-0.10
NV3	NS		1.00	0.33	0.68	-0.02
VCC3	NS	***		0.27	0.68	0.12
VSC3	NS	***	**		0.19	0.20
CT3	NS	***	***	*		0.36
TA3	NS	NS	NS	NS	NS	

Correlations auarta collection						
	MS4	NV4	VCC4	VSC4	CT4	TA4
MS4		0.09	0.10	-0.14	0.09	0.10
NV4	NS		1.00	-0.03	0.69	0.59
VCC4	NS	***		-0.09	0.69	0.60
VSC4	NS	NS	NS		-0.11	-0.14
CT4	NS	***	***	NS		0.93
TA4	NS	***	***	NS	***	
Correlations or collection						
	MS5	LV5	VCC5	VSC5	CT5	TA5
MS5		0.19	0.18	0.14	0.18	0.21
LV5	NS	-	1.00	0.11	0.69	0.60
VCC5	NS	***		0.05	0.71	0.61
VSC5	NS	NS	NS		-0.14	-0.16
CT5	NS	***	***	NS		0.92
TA5	NS	***	***	NS	***	-

*significant correlation by t test ($p < 0.05$)

**significant correlation by t test ($p < 0.01$)

*** significant correlation by t test ($p < 0.001$)

For the correlation results, the MS characteristic did not show significance with any of the others on any of the collection days, a statistic that allows us to state that the food efficiency test applied to select the best animals in this group was not harmed, not being influenced by the variations of MS.

Lopes (2011), looking for answers to the interference of physical variables in the quality of milk in dairy cows, evaluated DM during the day, but only applied correlation analyzes between granulometric characteristics, as according to the results of this study, variations in DM do not they have a direct correlation with consumption variables and eating behavior.

Among the characteristics of visits to the troughs (NV, VCC and VSC) there were significant correlations, mostly low, negative and positive, on all collection days, without a trend over the days.

The VSC variable was tested in correlations, to verify its influence on the other variables, which was not significant in most results. This may indicate that these visits to the trough without any food consumption are sporadic events, capable of being explained by the animal's own behavior of curiosity and, disputes for space, scratching, among others, because whenever the head reaches the trough the system will read the earring, regardless of the reason that took you to the location.

For CT and TA, except on the third day of collections, on the others, they were highly significant and positive, when correlated with visits, indicating that, the greater the number of visits, the greater the consumption and total feeding time tend to be. . It is also observed for the variables in question, that in the last two days of evaluation, the correlation values were very high, a fact that tends to stabilize as the days go by, due to the satiety of the cattle and fewer disputes over the trough.

These results can be compared with studies by Padiora et al. (2011), who found an increase in the number of visits to the trough as the number of daily treatments increased and, consequently, an increase in performance. The more animals go to the trough, the more time they spend feeding and gain weight.

Silva et al. (2017) concluded in their experiments, in a food efficiency test, with the Intergado electronic trough system, that CAR presents significant associations with characteristics of ingestive behavior of beef cattle in confinement, with more efficient animals for this measure visiting less frequently at the trough and spend less time feeding. This cannot be measured in the experiment in question, as it was not possible to obtain the number of visits to the trough and feeding time for each animal, individually, throughout the experiment.

Regression analyzes were also carried out for the characteristics of dry matter and total consumption, which presented parametric results, however, the $R^{2\ \text{results}}$ were low (less than 20%), both linear and quadratic, which does not allow explaining efficiently the behavior over time of the variables in question.

For particle size characteristics, after carrying out statistical analyzes on the data obtained through the PSPS, when comparing the averages of the time intervals (1 to 14), no significant results were found on any of the evaluation days, either for parametric analyses. as for non-parametric ones.

This fact allows us to state that the physical quality of the diet, measured through granulometry, remained the same throughout the day, during approximately twelve hours of evaluation, where the animals were more active. Therefore, the efficiency test carried out to select the best animals for CAR had no influence on the quality of the food provided to the animals, maintaining the quality and standard throughout the day.

Lopes (2011) mentions that, throughout the day, the largest fraction of the total food tends to be retained in the Imm sieve, if the animals eat the larger particles first, or in the 19 mm sieve, if they manage to select the smallest particles and the feed concentrate. In this work, none of these aforementioned trends were evident, as there were no significant differences between the 14 times where the granulometry of the total diet was measured.

The results found in this study corroborate Garcia (2009) who found that, the more times during the day the PSPS is measured, the better the data will be for analyzing the feeding behavior of dairy cows, and from that , make decisions. We must consider this report for any category of cattle production, as the fact that they are ruminants and have the ability to select food *is* general to all.

In studies by Marano et al. (2014), the authors also found no significant differences in the particle size profile of diets for confined beef cattle using the *GrowSafe* feeding system, also corroborating the results of this research.

Analyzes were also carried out for the average results of each collection day, within the sieve divisions (Table 11), where the results did not follow the values proposed as ideal by Heinrichs and Kononoff (2004).

Table 11 - Average values obtained in percentages among PSPS containers for each collection day.

Sieves / Treatments	Averages %			
	19mm	8mm	1mm	bottom
First Collection	31.41 [a]	23.51b	17.32c	26.12 [b]
Second Collection	19.31b	30.94 [to]	17.18 [b]	29.99 [to]
Third Collection	14.69c	28.06 [a]	13.92c	21.13b
Fourth Collection	12.33d	40.18 [to]	19.07c	26.17b
Fifth Collection	14.63d	36.68 [a]	19.69c	26.93b

*means followed by different letters in each line, differ from each other using the Kruskal-Wallis test at 5%

probability.

Heinrichs and Kononoff (2004), cite as ideal in the first sieve (19 mm) retention of 2 to 8% of particles, in the second (8 mm) 30 to 50%, in the third (1 mm) also 30 to 50% and the bottom tray <= 20%, with a forage:concentrate ratio approximate to that of the experiment in question (60:40), however, its recommendations are for dairy cow diets. When we consider this fact, tolerance levels and ideal values for a diet for beef cattle in confinement certainly change, but data for such comparisons are non-existent in the literature.

The results of Lopes (2011) for granulometry also do not follow the ideal standard proposed by the authors mentioned above, who carried out their studies on feed for dairy cows in Lisbon. The feed tested by the author in question presented values lower than those recommended as ideal on the middle sieves (8 and 1 mm).

Even so, considering the forage:concentrate ratio of the diet provided to the animals in question (60:40), based on the results obtained, these are considered good, due to the proportions of food retained at the two extremes of the PSPS, which must be smaller and the two middle compartments larger.

Custodio et al., (2016), testing different sources of forage such as roughage in confinement diets for beef cattle, with a roughage:concentrate ratio of approximately 25:75, concluded after passing the feed through the PSPS that the highest percentage of consumption of Animals confined throughout the day had medium and large particles (8 and 19 mm sieves). This fact has a positive aspect for the functioning of the rumen, depending on the feed and animal category used, as these are highly concentrated feeds.

In this experiment, we can observe that there was a strong tendency for most of the food to be deposited on the second sieve and the bottom (8 mm and tray), and the smallest part on the first and third sieve (19 and 1 mm).

Considering the differences in the roughage:concentrate ratio between the studies, these results corroborate the studies carried out by the author mentioned above.

CHAPTER 5

CONCLUSION

It is concluded that physical changes in grain size and dry matter in the diet throughout the day did not influence the consumption of feed by Senepol cattle.

It is emphasized that these results were found using three daily treatments, electronic and covered troughs. The results obtained under different conditions may not be similar, however, tests need to be carried out to be aware if there are variations.

REFERENCES

ALMEIDA, R. de. **The importance of routinely determining the dry matter of foods on the farm.** MilkPoint Technical Radar, Curitiba - PR, 2006. Available at: < https://www.milkpoint.com.br/radar-tecnico/nutricao/a- importancia-de-determinar-rotineiramente-a-materia-seca-dos- foods-on-the-farm-27695n.aspx >. Accessed on: 01/11/2018.

BRITO, F. **Adjustment in the dry matter (DM) of the diet of beef cattle.** Agroceres Multimix, 2017. Available at: < http://www.agroceresmultimix .com.br/blog/ajuste-na-materia-seca-ms- da-dieta-de-bovinos-de-corte/ >. Accessed on: 01/11/2018.

CARVALHO, TB de; ZEN, S. de. The beef cattle farming chain in Brazil: evolution and trends. **iPecege Magazine,** v. 3, no. 1, p. 85-99, 2017.

CASTILHOS, AM de. **Feed efficiency and performance of Nelore cattle selected for post-weaning weight.** 2009. 85f., Dissertation (Masters in Animal Science), Universidade Estadual Paulista. Faculty of Veterinary Medicine and Animal Science. Department of Animal Science. Botucatu - SP.

Center for Advanced Studies in Applied Economics (CEPEA). 2016. **Agribusiness GDP.** Available at: < http://cepea.esalq.usp.br/pib/ >. Accessed on: 06 Jan. 2017.

COELHO, RM **Effects of dry matter concentration and the use of bacterial-enzymatic inoculant, in Tifton 85 (Cynodon spp.) silage, on nutrient digestion, ruminal parameters and ingestive behavior in growing beef steers.** 2002. 143 f. University of São Paulo - USP, Piracicaba - SP.

COSTA JÚNIOR, JR et al. Factors influencing mixing quality in confinement diets.

Anais... X SEZUS, São Luís de Montes Belos - GO, 2016.

CUSTODIO, SAS **Performance, selective consumption of feed particles and ingestive behavior of beef cattle.** 2016. 82 f. Dissertation (Master's degree in Animal Science), Postgraduate Program in Animal Science at the Instituto Federal Goiano, Rio Verde - GO.

CUSTODIO, SAS et al. Selection of particles in the diet of beef cattle in confinement fed with different forages and housed in stalls

individual or collective. **Journal Animal Behavior Biometeorol,** v. 4, no. 2, p. 55-64, 2016.
https://doi.org/10.14269/2318-1265/iabb.v4n2p55-64

DAMASCENO, JC; JUNIOR, FB; TARGA, LA Behavioral responses of Holstein cows with access to constant or limited shade. **Brazilian Agricultural Research,** v. 34, p. 709-715, 1999.
https://doi.org/10.1590/S0100-204X 1999000400024

DETMANN, E. et al. Consumption of neutral detergent fiber by cattle in confinement. **Brazilian Journal of Animal Science.** V.32, n.6, p. 1763-1777, 2003.
https://doi.Org/1 0.1590/S1516- 35982003000700027

DEVRIES, TJ et al. Repeated ruminal acidosis challenges in lactating dairy cows at high to low risk for developing acidosis: feed sorting. **Journal of Dairy Science.** V. 91 n. 10, p. 3958-67, 2008.
https://doi.org/10.3168/ids.2008-1347

EUCLIDES FILHO, K. Evolution of genetic improvement of beef cattle in Brazil. **Review Ceres,** Viçosa, v. 56, no. 5, p. 620-626, Sep/Oct, 2009.

GARCIA, AR **Usage of the Penn State Forage Separator for evaluating particle size of TMRs.** Dairy Science Department, California Polytechnic State University, San Luis Obispo, USA, 2009.

GOMES, Rodrigo da Costa; MENDES, Egleu DM **Procedures for measuring individual food consumption in beef cattle.**
ResearchGate, 2013. Available at: < https://www.researchgate.net/publication 7259099649 Procedures for measuring individual feed intake in beef cattle?enrichld=rgreq-8f62d9d75ea62e566a7cca9de85d 8b3- OTYQOTtBUzoxMDE yMjk5NzU4MzQ2MzlAMTQwMTE0NjUxN TgzNw%3D%3D&el =lx 2& esc=publicationCoverPdf > Accessed on: 03/12/2018.

GOULART, R. **Efficiency in mixing feeds.** AG - The Creator's Magazine.

Ed.162. 2012.

GROESBECK, CN et al. Effects of salt particle size and sample preparation on results of mixer-efficiency testing. **Annals...** Porkinformation, Kansas, p. 177- 188. 2006.

HEINRICHS, Jude; KONONOFF, Paul. **Evaluating particle size of forages an TMRs using the New Penn State Forage Particle Separator.** 2004. Department of Dairy and Animal Science, Pennsylvania State University, USA.

KRAUSE, KM; OETZEL, GR Understanding and preventing subacute ruminal acidosis in dairy herds: a review. **Animal Feed Science and Technology.** V. 126, n. 3 - 4, p. 215 - 236, 2006.

LAGE, BFC **Relationship between feeding behavior and temperament with residual feed intake in Nelore steers.** 2013. 51 f. Dissertation (Master's degree in Animal Science), Postgraduate Program in Animal Science at the Federal University of Vales do Jequitinhonha and Mucuri. Diamantina - MG.

LAZARINI, VF; GAI, VF; FAGUNDES, RS Chemical composition of the diet in relation to beating time. **Cultivating Knowledge,** v.7, nl, p. 102-110, 2014.

LOPES, ARMF **Influence of the mixing process on the physical and chemical composition of single feed for dairy cows.** 2011. 77 f. Dissertation (Zootechnical Engineering), Technical University of Lisbon. Lisbon, Portugal.

LUIZ, FP Ingestive behavior of young cattle in confinement receiving rations with ingredients from plants with C3 or C4 photosynthetic cycle. **Proceedings...** 3rd FATEC Scientific and Technological Conference in Botucatu, Botucatu- SP, 2014 .

MARANO, WA et al. Particle size profile in chemical composition of leftovers from the diet of Nelore heifers in an automatic feeding system. **Proceedings...** 8th Interinstitutional Congress of Scientific Initiation - CIIC, Campinas - SP, 2014.

MEDEIROS, SR de; et al. **Precision livestock tools aimed at the nutrition of beef cattle.** 22 p. 2016.

MENDES, EDM; CAMPOS, MM Feed efficiency in beef cattle. **Agricultural Report,** Belo Horizonte, v. 37, no. 292, p. 28-38, 2016.

MORAES, GF et al. **Use of residual food consumption for the genetic improvement of beef cattle.** Embrapa Cerrados, Planaltina - DF, 2016.

NEUMANN, M. et al. Performance, dry matter digestibility and ingestive behavior of Holstein steers fed different diets in confinement. **Semina: Agricultural Sciences,** Londrina - PR, v. 36, no. 3, p. 1623-1632, 2015.

NKRUMAH, JD et al. Relationships of feedlot feed efficiency, performance, and feeding behavior with metabolic rate, methane production, and energy partitioning in beef cattle. **Journal of Animal Science,** vol. 84, p.145-153, 2006. https://doi.org/10.2527/2006.841145x

PAZDIORA, RD et al. Effects of the frequency of roughage and concentrate supply on the ingestive behavior of cows and heifers in confinement. **Brazilian Journal of Animal Science,** v.40, n.10, p. 2244-2251, 2011. https://doi.org/10.1590/S1516-35982011001000026

PEREIRA, JCC **Genetic improvement applied to animal production.** 4th ed. Belo Horizonte: FEPMVZ Editora, 2014. 758 p.

PEREIRA, IC **Meta-analytic study of fluctuations in dry mass intake on performance, ingestive behavior and ruminal health of cattle confined on high-concentrate diets.** 2016. 79 f. Dissertation (Zootechnics). São Paulo State University - UNESP, Botucatu - SP.

PHILLIPS, C.J.C.; RIND, MI The effects of social dominance on the production and behavior of grazing dairy cows offered forage supplements. **Journal of Dairy Science,** vol. 85, no. 1, p. 51-59, 2002. https://doi.org/10.3168/jds.S0022-0302(02)7 4052-6

QUINTILIANO, MHE; COSTA, MJR Paranhos. Rational management of beef cattle in confinement: Productivity and animal welfare. **Annals...** IV SINEBOV, Seropédica, RJ, 2006.

ROTGER, A. et al. Effects of dietary nonstructural carbohydrates and protein sources on feeding behavior of tethered heifers fed high-concentrate diets. **Journal of Animal Science,** v.84, p. 1197-1204, 2006. https://doi.org/10.2527/2006.8451197x

RUSTOMO, B. et al. Effects of rumen acid load from feed and forage particle size on ruminal pH and dry matter intake in the lactating dairy cow. **Journal Dairy Science,** no. 89, p. 4758^4768, 2006. https://doi.org/10.3168/jds .S0022-0302(06)72525-5

RUTTER, SM **Behavior of sheep and goats.** In. Jensen, P. (ed.) The ethology of domestic animals: an introductory text. CAB International, Wallingford, 2002, p.

145-158.
https ://doi.org/10.1079/9780851996028.0145

SCHUTZ, JS; JJ et al. Effect of feeding frequency on feedlot steer performance. **The Professional Animal Scientist,** no. 27, p.14 - 18, 2011.
https://doi.org/10.15232/S1080-7446(15)30439-3

SCHWARTZKOPF-GENSWEIN, KS et al. Effect of bunk management on feeding behavior, ruminal acidosis and performance of feedlot cattle: A review. **Journal of Animal Science,** Savoy, vol. 81, p. 149-158, 2003.

SCHWARTZKOPF-GENSWEIN, KS, ATWOOD, S., McALLISTER, TA Relationships between bunk frequency, intake and performance of steers and heifers on varying feeding regimes, **Applied Animal Behavior Science,** v. 76, p. 179-188, 2002.
https://doi.org/10.1016/S0168-1591 (02)00009-6

SIGNORETTI, RD et al. Consumption and apparent digestibility in Holstein calves fed diets containing different roughage levels.
Brazilian Journal of Animal Science, v.28, nl, p. 169-177, 1999.
https://doi.org/10.1590/S1516-35981999000100025

SILVA, AM et al. **Association between feed efficiency and ingestive behavior of Nelore bulls confined in collective stalls equipped with electronic troughs.** 2017.

SILVA, J. da. et al. Feedlot performance, feeding behavior and rumen morphometrics of Nellore cttle submitted to different feeding frequencies.
Scientia Agrícola, v.75, n.2, p.121-128, 2018.
https://doi.org/10.1590/1678-992x-2016-0335

SILVA, J. **Feeding frequency on ingestive behavior, starch digestibility and consumption fluctuation in confined Nellore cattle.** 2014a. 44 f. Dissertation (Animal Science and Technology), Universidade Estadual Paulista, Dracena - MG.

SILVA, SRC **Performance and carcass characteristics of Nelore cattle with different consumption habits in confinement.** 2014b. 50 f.
Dissertation (Masters in Animal Science), Federal University of Goiás, Goiânia - GO.

SOVA, AD, LEBLANC, SJ, MCBRIDE, BW, and DEVRIES, TJ Accuracy and precision of total mixed rations fed on commercial dairy farms. **Journal of Dairy Science,** vol. 97, no. 1, p. 562-571, 2014.
https://doi.org/10.3168/jds.2013-6951

STIEVEN, ICB **Relationships between residual food consumption and hematological profile, stress and ingestive behavior in Purunã cattle.** 2012. 81 f. Dissertation (Master in Veterinary Sciences), Agricultural Sciences Sector, Postgraduate Program in Veterinary Sciences. Federal University of Paraná Curitiba - PR.

TAKIGAWA, TMY **Trough management: its importance and how to do it.** Premix Technical Article, 8th [ed]., 2015.

VALADARES FILHO, S. de C. et al. Perspectives on the use of indicators to estimate individual consumption of cattle fed in groups. **Proceedings... Symposium of** [the] 43rd Annual Meeting of the SBZ - João Pessoa - PB, 2006.

Printed by Books on Demand GmbH, Norderstedt / Germany